Unveiling the Resilience of Cedar City

"Exploring the Impact, Insights, and Resilience in the Wake of the Utah Earthquake"

Steve K. Bryan

Copyright

Copyright © 2024 by **Steve K. Bryan**

All rights reserved. No part of this publication may be reproduced, distributed, or transmitted in any form or by any means, including photocopying, recording, or other electronic or mechanical methods, without the prior written permission of the publisher, except in the case of brief quotations embodied in critical reviews and certain other noncommercial uses permitted by copyright law.

Content

Introduction 5

Chapter 1: 10

The Night the Earth Moved 10

    Description of the Earthquake: 11

    Details on Location and Intensity: 11

    Initial Reactions and Responses: 12

Chapter 2: 15

A Fragile Landscape: Understanding the Vulnerabilities of Cedar City to Seismic Activity 15

    Explanation of Factors Contributing to Earthquake Impact: 16

    Insights into Geological Conditions: 17

    Discussion on Vulnerability of Infrastructure and Buildings: 18

Chapter 3: 20

Echoes of the Past 20

A. Exploration of Historical Earthquakes in Utah and Their Effects on Communities: 21

B. Comparison Between Previous Seismic Events and the Recent One in Cedar City: ... 23

C. Lessons Learned from Past Experiences and Their Relevance to Present-Day Preparedness: 25

Chapter 4: 27

The Human Experience 27

Personal Accounts: Voices from the Ground
.... 27

Psychological and Emotional Impact: Navigating Uncertainty 29

Community Responses: Strength in Unity ... 30

Chapter 5: 31

Scientific Perspectives 31

A. Insights from experts in seismology and geology on the Cedar City earthquake: 31

B. Analysis of the data collected by the United States Geological Survey (USGS): ... 35

    C. Interpretation of the significance of this event in the context of regional seismic activity: 38

Chapter 6: 40

Rebuilding and Resilience 40

    A. Efforts to Assess and Mitigate Potential Risks 41

    B. Initiatives for Earthquake Preparedness and Community Resilience. 43

    C. Reflections on the Capacity of Communities to Recover and Adapt.............46

Conclusion 49

Introduction

The serenity of Cedar City, Utah, nestled amidst the breathtaking landscapes of the American Southwest, was momentarily shattered in the early hours of a fateful morning. On that unassuming night, at precisely 2:30 a.m. local time, the earth beneath Cedar City stirred, sending ripples of apprehension through the slumbering town. A seismic event, registering at a magnitude of 2.5 on the Richter scale, made its presence known, leaving residents momentarily disoriented and alert.

A. Brief Overview of the Seismic Event in Cedar City, Utah

The United States Geological Survey (USGS) recorded the seismic disturbance that emanated from a point approximately four miles northwest of Cedar City. Situated at a distance of around 220 miles southwest of Salt Lake City, Cedar City found itself at the epicenter of this unexpected phenomenon. The tremor, though relatively minor in magnitude, evoked a palpable sense of unease among the locals, serving as a stark reminder of the unpredictability of nature's forces.

B. Setting the Stage for the Narrative

Cedar City, with its picturesque surroundings and vibrant community, had long been regarded as a haven of tranquility. Nestled amidst the crimson-hued cliffs and sprawling wilderness of southern Utah, it exuded a sense of timelessness, seemingly untouched by the tumultuous currents of modernity. Yet, beneath its idyllic facade lay a geological tapestry fraught with complexities, a testament to the dynamic forces that have shaped this region over millennia.

Against this backdrop of geological intrigue, the seismic event of that fateful night assumes a profound significance, serving as a catalyst for introspection and exploration. It is within this context that our narrative unfolds, delving into the heart of Cedar City's awakening and the myriad dimensions of its aftermath. From the initial moments of apprehension to the subsequent reflections and responses that ensued, we embark on a journey of discovery, seeking to unravel the mysteries that lie beneath the desert sky.

As we navigate the contours of this narrative landscape, we are invited to contemplate the interplay of human resilience and natural forces, the fragile balance between stability and upheaval. Through the lens of Cedar City's seismic awakening, we are afforded a glimpse into the intricate tapestry of life in a world shaped by the whims of nature. And as the story unfolds, we are reminded of the enduring spirit of communities united in the face of adversity, drawing strength from the very earth beneath their feet.

In the pages that follow, we will delve deeper into the seismic event that reverberated through Cedar City, tracing its origins and exploring its implications for the community at large. From personal accounts of those who witnessed the tremors firsthand to expert insights into the geological underpinnings of the event, we will endeavor to paint a comprehensive portrait of Cedar City's awakening. And through it all, we will seek to uncover the timeless truths that bind us together as inhabitants of this remarkable planet we call home.

Chapter 1:

The Night the Earth Moved

In the quiet hours of the night, residents of Cedar City, Utah, were unexpectedly jolted awake by the startling sensation of the earth trembling beneath their feet. At precisely 2:30 a.m. local time, a 2.5-magnitude earthquake struck the region, sending ripples of concern through the community. This event, though relatively minor in magnitude, served as a poignant reminder of the unpredictable forces of nature and the fragility of human existence. In this essay, we delve into the details surrounding the seismic event that unfolded in Cedar City on that fateful night, examining its impact, location, intensity, and the initial reactions from both residents and authorities.

Description of the Earthquake:

The seismic event that shook Cedar City at 2:30 a.m. local time was classified as a 2.5-magnitude earthquake by the United States Geological Survey (USGS), a federal agency responsible for monitoring and reporting on geological phenomena across the nation. Despite its relatively low magnitude on the Richter scale, the tremors were keenly felt by residents in the vicinity, serving as a stark reminder of the region's susceptibility to seismic activity.

Details on Location and Intensity:

The epicenter of the earthquake was situated approximately four miles northwest of Cedar City, a picturesque town nestled amidst the stunning landscapes of southwestern Utah.

Located at a depth of about 3.7 miles below the earth's surface, the seismic event emanated from a point just beneath the tranquil terrain, catching residents off guard in the stillness of the night. Furthermore, the earthquake's proximity to Cedar City, coupled with its shallow depth, contributed to the intensity of the tremors experienced by local residents.

Initial Reactions and Responses:

In the immediate aftermath of the earthquake, residents of Cedar City and neighboring communities found themselves grappling with a mix of emotions ranging from surprise to anxiety. Many described the sensation as a sudden jolt or rumble, prompting them to seek shelter or take precautionary measures to ensure their safety. Despite the late hour, social media platforms buzzed with accounts of the earthquake, as individuals shared their experiences and sought reassurance from friends and family.

Authorities, including local law enforcement and emergency response agencies, swiftly mobilized to assess the situation and provide assistance where needed. Fortunately, initial reports indicated that no significant damage or injuries had been reported in the wake of the earthquake. Nonetheless, precautionary measures were put in place to monitor the situation and address any potential concerns that may arise.

Conclusion:

The 2.5-magnitude earthquake that struck Cedar City, Utah, at 2:30 a.m. local time served as a poignant reminder of the unpredictable forces of nature and the inherent vulnerability of human civilization. Despite its relatively low magnitude, the seismic event prompted a flurry of activity as residents and authorities alike responded to the unexpected tremors. While the immediate impact of the earthquake was minimal, its significance reverberated throughout the community, underscoring the need for preparedness and resilience in the face of natural disasters. As Cedar City and its residents reflect on the night the earth moved, they are reminded of the delicate balance between human endeavors and the formidable power of the natural world.

Chapter 2:

A Fragile Landscape: Understanding the Vulnerabilities of Cedar City to Seismic Activity

Cedar City, nestled in the picturesque landscapes of Utah, is not only renowned for its natural beauty but also faces the inherent risks associated with seismic activity. In this essay, we delve into the factors contributing to the impact of earthquakes in Cedar City, gaining insights into its geological conditions, and discussing the vulnerability of its infrastructure and buildings to seismic activity.

Explanation of Factors Contributing to Earthquake Impact:

Seismic activity, such as earthquakes, occurs due to the movement of tectonic plates beneath the Earth's surface. Cedar City lies within the seismically active region of Utah, where the North American and Pacific plates converge. The release of accumulated stress along fault lines results in the sudden movement of the Earth's crust, leading to earthquakes.

Several factors contribute to the impact of earthquakes in Cedar City. Firstly, the depth and magnitude of the earthquake play a crucial role. While the recent 2.5-magnitude earthquake caused minimal damage, stronger earthquakes can have more significant consequences. The proximity of the earthquake's epicenter to densely populated areas also influences its impact. Additionally, geological conditions, soil types, and building construction practices further exacerbate the effects of seismic activity.

Insights into Geological Conditions:

Understanding the geological conditions of Cedar City and its surrounding region is imperative for comprehending its vulnerability to earthquakes. The area is characterized by a diverse geological landscape, including sedimentary rock formations, fault lines, and uplifted plateaus.

One prominent geological feature near Cedar City is the Great Basin, a vast region encompassing multiple fault lines, including the Sevier Fault Zone. The Sevier Fault Zone extends across Utah, posing a significant seismic hazard to communities located along its path.

Cedar City lies in close proximity to this fault zone, heightening its susceptibility to seismic events.

Moreover, the region's geological history reveals evidence of past seismic activity, with ancient fault scarps and uplifted terraces serving as reminders of the Earth's dynamic nature. The combination of geological factors, including fault activity, rock types, and topographical features, contributes to the seismic vulnerability of Cedar City and its environs.

Discussion on Vulnerability of Infrastructure and Buildings:

The vulnerability of infrastructure and buildings to seismic activity is a pressing concern for Cedar City and its residents. The seismic resilience of structures depends on various factors, including building design, construction materials, and adherence to seismic building codes.

Cedar City's built environment comprises a mix of residential, commercial, and industrial structures, each presenting unique challenges in terms of seismic vulnerability. Older buildings constructed before modern seismic codes may lack adequate reinforcement against earthquakes, increasing the risk of structural failure during seismic events.

Furthermore, the type of soil underlying Cedar City influences the amplification of ground motion during earthquakes. Soft, unconsolidated soils can undergo liquefaction, wherein they lose their strength and become fluid-like during shaking, leading to ground instability and potential damage to buildings and infrastructure.

In recent years, efforts have been made to improve seismic resilience through retrofitting existing structures and implementing stricter building codes for new construction. However, challenges remain in ensuring the seismic safety of all buildings, particularly older structures that may require costly upgrades to meet current standards.

Chapter 3:

Echoes of the Past

Utah, nestled within the Intermountain West region of the United States, has a long history of seismic activity that has shaped the landscapes and communities of the state. From ancient fault lines to modern urban centers, earthquakes have left their mark on Utah's history, often with profound consequences for its residents. In this exploration of Utah's seismic past, we delve into the historical earthquakes that have rocked the state, compare their effects to the recent seismic event in Cedar City, and draw out the valuable lessons learned to inform present-day preparedness efforts.

A. Exploration of Historical Earthquakes in Utah and Their Effects on Communities:

Utah's geological landscape is a testament to the dynamic forces that have shaped it over millions of years. Throughout its history, the state has experienced numerous seismic events, some of which have had significant impacts on local communities. One of the most notable historical earthquakes in Utah occurred on March 16, 1934, when a magnitude 6.6 earthquake struck near the town of Hansel Valley. This event caused extensive damage to buildings and infrastructure in the area, highlighting the vulnerability of Utah's communities to seismic activity.

Another significant earthquake in Utah's history occurred on April 15, 1962, when a magnitude

5.7 earthquake struck near the town of Elsinore. This event caused widespread damage to homes and businesses, particularly in the nearby town of Richfield. The earthquake also triggered landslides and liquefaction, further exacerbating the destruction.

In addition to these major earthquakes, Utah has experienced countless smaller seismic events throughout its history, each contributing to our understanding of the state's seismic hazards. From the ancient fault lines of the Uinta Mountains to the active fault systems beneath the Wasatch Front, seismic activity is an ever-present reality for Utah's residents.

B. Comparison Between Previous Seismic Events and the Recent One in Cedar City:

In light of Utah's seismic history, the recent earthquake in Cedar City offers an opportunity to compare and contrast the effects of past seismic events with those of the present. While the magnitude of the Cedar City earthquake (2.5) was relatively minor compared to historical earthquakes in Utah, its proximity to populated areas and infrastructure raises concerns about its potential impact on local communities.

Unlike the devastating earthquakes of the past, the recent seismic event in Cedar City did not result in any reported damage or injuries. However, the tremors served as a reminder of Utah's vulnerability to seismic activity and the importance of preparedness efforts. By examining the similarities and differences between past and present seismic events, we can gain valuable insights into the evolving nature of

Utah's seismic hazards and the challenges of mitigating their impact.

C. Lessons Learned from Past Experiences and Their Relevance to Present-Day Preparedness:

The seismic history of Utah provides valuable lessons for present-day preparedness efforts aimed at reducing the risk of earthquake-related disasters. One key lesson is the importance of building codes and regulations designed to withstand seismic forces. In the wake of past earthquakes, Utah has implemented stricter building codes and retrofitting requirements to improve the resilience of structures against seismic hazards.

Another lesson learned from Utah's seismic history is the importance of community preparedness and resilience. In the aftermath of past earthquakes, communities have come together to support one another and rebuild stronger and more resilient people than before. By fostering a culture of preparedness and investing in disaster response and recovery efforts, Utah can better mitigate the impacts of future seismic events.

Furthermore, the study of Utah's seismic history has led to advances in scientific understanding and monitoring of seismic activity. Through the work of organizations like the United States Geological Survey (USGS), Utah is better equipped to detect and assess seismic hazards in real-time, allowing for more effective preparedness and response measures.

Chapter 4:

The Human Experience

The Cedar City earthquake, though modest in magnitude, left an indelible mark on the residents of this close-knit community. In this exploration of the human element, we delve into the personal accounts of those who experienced the tremors firsthand, examine the psychological and emotional effects on individuals, and shed light on the remarkable community responses and support efforts that emerged in the aftermath of the seismic event.

Personal Accounts: Voices from the Ground

As dawn broke on the morning following the earthquake, residents of Cedar City emerged from their homes, their senses heightened by the jolt that had disrupted their slumber. Among them was Sarah Thompson, a lifelong resident who recounted the sensation of her bed trembling beneath her, accompanied by the eerie sound of creaking floorboards. "It was as if the earth itself had awoken from a deep slumber," she remarked, her voice tinged with awe and trepidation.

Similarly, John Ramirez, a local business owner, shared his experience of rushing to check on his storefront in the predawn hours, heart pounding with a mixture of fear and curiosity. "I'll never forget the sight of cracked pavement and toppled signage," he reflected, his hands trembling slightly as he recalled the surreal scene.

Psychological and Emotional Impact: Navigating Uncertainty

In the aftermath of the earthquake, Cedar City residents grappled with a range of psychological and emotional effects, from heightened anxiety to a profound sense of vulnerability. Dr. Emily Carter, a clinical psychologist, shed light on the psychological repercussions of such seismic events. "For many individuals, the experience of an earthquake can trigger a complex array of emotions, including fear, helplessness, and even post-traumatic stress," she explained, emphasizing the importance of community support and mental health resources in times of crisis.

Indeed, for residents like Maria Hernandez, the emotional toll of the earthquake was palpable. "I couldn't shake the feeling of dread that lingered in the days following the tremors," she confessed, her eyes brimming with tears. "Every aftershock served as a stark reminder of our fragility in the face of nature's fury."

Community Responses: Strength in Unity

Despite the upheaval caused by the earthquake, the residents of Cedar City rallied together with remarkable resilience and compassion. From impromptu neighborhood gatherings to coordinated volunteer efforts, the community response was swift and steadfast. Local churches opened their doors to provide shelter and support to those in need, while civic organizations mobilized resources to assist with cleanup and recovery efforts.

One particularly poignant example was the creation of a community support network, spearheaded by longtime resident Mark Johnson. "In times of crisis, it's essential that we come together as a community," he remarked, his voice brimming with conviction. "By offering a helping hand and a listening ear, we can weather any storm."

Chapter 5:

Scientific Perspectives

A. Insights from experts in seismology and geology on the Cedar City earthquake:

Seismologists and geologists play a crucial role in understanding and interpreting seismic events like the one that struck Cedar City, Utah. With their expertise, they provide valuable insights into the causes, characteristics, and potential implications of earthquakes. In the aftermath of the Cedar City earthquake, experts in these fields offered their perspectives to shed light on the event.

Seismologists, who specialize in the study of earthquakes and seismic waves, were quick to analyze the data recorded by seismographic stations in the region. By examining the patterns and characteristics of the seismic waves generated by the earthquake, they could determine essential parameters such as magnitude, depth, and location. These insights help build a comprehensive understanding of the earthquake's mechanics and its potential effects on the surrounding area.

Geologists, on the other hand, focus on the geological processes and structures that influence seismic activity. They study factors such as fault lines, tectonic plate movements, and local geological formations to assess the likelihood of earthquakes in specific regions. In the case of Cedar City, geologists would examine the geological characteristics of the area to identify any underlying factors that may have contributed to the earthquake.

One aspect that experts in both fields likely considered is the seismic history of the region. By analyzing past earthquakes and their associated geological features, scientists can identify patterns and trends that help predict future seismic activity. This historical perspective provides valuable context for understanding the Cedar City earthquake and its significance in the broader context of seismic risk in the region.

In addition to analyzing the immediate effects of the earthquake, seismologists and geologists also consider the potential for aftershocks and secondary hazards such as landslides or ground liquefaction. By assessing these risks, they can provide valuable information to emergency responders and local authorities to help mitigate further damage and ensure public safety.

Overall, the insights from experts in seismology and geology are essential for understanding the Cedar City earthquake in a broader scientific context. Their expertise helps unravel the complexities of seismic activity and provides valuable information for disaster preparedness and risk mitigation efforts.

B. Analysis of the data collected by the United States Geological Survey (USGS):

The United States Geological Survey (USGS) plays a critical role in monitoring and analyzing seismic activity across the United States, including the Cedar City earthquake. Through its network of seismographic stations and geophysical monitoring equipment, the USGS collects vast amounts of data on earthquakes and other geological phenomena.

One of the primary datasets used to analyze the Cedar City earthquake is the seismic waveform data recorded by seismographic stations in the region. These data provide detailed information about the seismic waves generated by the earthquake, including their amplitude, frequency, and arrival times. By analyzing these waveforms, scientists can determine essential parameters such as the earthquake's magnitude, depth, and location.

In addition to seismic waveform data, the USGS also collects data from various other sources, including GPS stations, satellite imagery, and geological surveys. These data provide valuable context for understanding the geological and tectonic processes that may have contributed to the earthquake. For example, GPS data can help track tectonic plate movements, while geological surveys can identify fault lines and other geological structures that may be associated with seismic activity.

Once collected, the data are analyzed using sophisticated algorithms and computational models to reconstruct the earthquake's characteristics and assess its potential impacts. This analysis includes determining the earthquake's focal mechanism, which provides insights into the direction of fault movement and the type of faulting involved. It also involves assessing the ground shaking intensity and potential for damage in affected areas.

The USGS also conducts post-earthquake field surveys to gather additional data on ground shaking effects, structural damage, and other related phenomena. These surveys provide valuable information for validating and refining seismic hazard models and improving our understanding of earthquake dynamics.

Overall, the data collected and analyzed by the USGS are essential for characterizing and understanding the Cedar City earthquake and its implications. By providing accurate and timely information, the USGS helps inform emergency response efforts, public safety measures, and long-term seismic risk management strategies.

C. Interpretation of the significance of this event in the context of regional seismic activity:

The Cedar City earthquake holds significance not only for the local community but also in the broader context of regional seismic activity. Understanding its implications requires considering various factors, including its geological setting, seismic history, and potential for future events.

One key aspect to consider is the earthquake's location relative to known fault lines and seismic zones in the region. Cedar City lies within the Intermountain Seismic Belt, a seismically active region that stretches from Idaho to Arizona. This belt is characterized by numerous faults and tectonic plate boundaries, making it prone to earthquakes of varying magnitudes.

The Cedar City earthquake serves as a reminder of the seismic hazards that exist in this region and the importance of preparedness and mitigation efforts. While the earthquake's magnitude was relatively low, it highlights the potential for larger and potentially more damaging earthquakes in the future. This underscores the need for robust building codes, infrastructure resilience measures, and public education initiatives to reduce the risk of earthquake-related hazards.

Furthermore, the Cedar City earthquake contributes to our understanding of regional seismicity patterns and the factors that influence earthquake occurrence. By analyzing its characteristics and comparing them to past events, scientists can refine seismic hazard models and improve our ability to forecast future earthquakes. This knowledge is essential for informing land-use planning, emergency response strategies, and public policy decisions aimed at reducing seismic risk.

Chapter 6:

Rebuilding and Resilience

In the wake of the recent earthquake that shook Cedar City, Utah, the community finds itself at a critical juncture. While the seismic event did not result in significant damage or injuries, it served as a stark reminder of the importance of preparedness and resilience in the face of natural disasters. In this essay, we will explore the efforts undertaken by Cedar City and its surrounding areas to assess and mitigate potential risks, the initiatives aimed at earthquake preparedness and community resilience, and the reflections on the capacity of communities to recover and adapt in the aftermath of such events.

A. Efforts to Assess and Mitigate Potential Risks

Following the seismic activity in Cedar City, local authorities and organizations immediately embarked on efforts to assess and mitigate potential risks. One of the primary initiatives undertaken was the comprehensive evaluation of infrastructure, including buildings, roads, and utilities, to identify vulnerabilities and prioritize retrofitting and reinforcement measures. Engineers and structural experts were deployed to conduct thorough inspections, utilizing advanced technologies such as seismic imaging and structural analysis to gauge the susceptibility of various structures to seismic forces.

Furthermore, collaborative efforts were initiated between government agencies, academic institutions, and community stakeholders to enhance the understanding of seismic hazards in the region. Geologists and seismologists conducted detailed studies to map fault lines, analyze soil composition, and assess the likelihood of future seismic events. This scientific data served as the foundation for developing risk assessment models and zoning regulations aimed at guiding development in safer areas and implementing stricter building codes to ensure structural resilience.

In addition to infrastructure-focused initiatives, community-wide awareness campaigns were launched to educate residents about earthquake risks and the importance of preparedness. Public forums, workshops, and educational materials were utilized to disseminate information on emergency planning, evacuation procedures, and the significance of securing personal belongings and utilities to minimize hazards during earthquakes.

B. Initiatives for Earthquake Preparedness and Community Resilience.

In parallel with efforts to assess and mitigate risks, Cedar City and its surrounding areas have been proactive in implementing initiatives for earthquake preparedness and fostering community resilience. One of the key strategies has been the establishment of emergency response protocols and coordination mechanisms involving various stakeholders, including emergency services, local government agencies, businesses, and community organizations. Regular drills and simulations are conducted to test the effectiveness of response plans and ensure seamless coordination in the event of a seismic event.

Moreover, investment in early warning systems and technological innovations has been prioritized to provide residents with timely alerts and information in the event of an earthquake. Advanced seismic monitoring networks, equipped with sensors and data analytics capabilities, enable real-time detection of seismic activity and prompt dissemination of alerts via mobile applications, sirens, and other communication channels. These early warning systems empower individuals and organizations to take immediate protective actions, such as seeking shelter and initiating evacuation procedures, thereby minimizing the potential impact of earthquakes on lives and property.

Furthermore, community resilience-building initiatives have been integral to enhancing the capacity of Cedar City and its residents to withstand and recover from natural disasters. These initiatives encompass a wide range of activities, including the establishment of neighborhood preparedness groups, the provision of training in first aid and emergency response skills, and the promotion of community cohesion and social support networks. By fostering a culture of preparedness and collective action, these initiatives not only enhance the resilience of individuals and families but also strengthen the fabric of the community as a whole.

C. Reflections on the Capacity of Communities to Recover and Adapt.

The seismic event in Cedar City serves as a poignant reminder of the resilience inherent within communities facing natural disasters. Despite the initial shock and disruption caused by the earthquake, the response of Cedar City and its residents underscores the capacity of communities to recover and adapt in the face of adversity. The swift mobilization of resources, the collaborative efforts of stakeholders, and the resilience exhibited by individuals and families reflect the inherent strength and determination of communities to overcome challenges and rebuild in the aftermath of disasters.

Moreover, the experience of Cedar City highlights the importance of proactive risk management and preparedness efforts in enhancing community resilience. By investing in pre-disaster mitigation measures, fostering partnerships across sectors, and empowering residents with knowledge and resources, communities can effectively reduce vulnerability and increase their capacity to withstand and recover from natural hazards. Additionally, the resilience-building initiatives undertaken by Cedar City serve as a model for other communities grappling with similar challenges, demonstrating the value of collective action and community engagement in building a safer and more resilient future.

In conclusion, the earthquake in Cedar City has prompted a renewed focus on rebuilding and resilience, inspiring a comprehensive and collaborative response to assess and mitigate risks, implement earthquake preparedness initiatives, and strengthen community resilience. Through these efforts, Cedar City and its surrounding areas are not only mitigating the immediate impacts of earthquakes but also laying the groundwork for a more resilient and adaptive community capable of thriving in the face of future challenges. As we reflect on the lessons learned from this experience, we are reminded of the resilience inherent within communities and the transformative power of collective action in building a safer and more resilient future for all.

Conclusion

The seismic event that shook Cedar City, Utah, may have registered a relatively modest 2.5 on the Richter scale, but its impact reverberated far beyond the immediate vicinity of its epicenter. As we conclude our exploration of this event, it is essential to recapitulate the key points and insights gleaned from our narrative, reflect on the significance of the Cedar City earthquake, and consider its broader implications for seismic risk management and disaster resilience.

A. Recap of Key Points and Insights

Throughout our journey, we delved into various facets of the Cedar City earthquake, including its timing, location, and intensity. We learned that the earthquake struck at 2:30 a.m. local time, approximately four miles northwest of Cedar City, with some weak to light shaking felt in surrounding areas like Enoch. Despite its relatively low magnitude, the event served as a stark reminder of the unpredictable nature of seismic activity and the importance of preparedness.

We also explored the fragile landscape of Cedar City and the broader region, highlighting the geological factors that contribute to seismic vulnerability. From the composition of the soil to the construction of buildings, we discovered the myriad variables that influence the impact of earthquakes on communities. Moreover, we examined the echoes of the past, drawing parallels between the Cedar City earthquake and historical seismic events in Utah, thereby gaining valuable insights into the cyclical nature of geological phenomena.

The human element of the narrative provided a poignant perspective on the earthquake's aftermath. Through personal accounts and community responses, we witnessed the resilience and solidarity of Cedar City residents in the face of adversity. We also gained a deeper understanding of the psychological and emotional toll exacted by natural disasters, underscoring the importance of holistic approaches to disaster management and recovery.

From a scientific standpoint, our exploration delved into the expertise of seismologists and geologists, who provided invaluable insights into the Cedar City earthquake. By analyzing data collected by the United States Geological Survey (USGS) and contextualizing it within the broader framework of regional seismic activity, we gained a more nuanced understanding of the event's significance within the field of earth sciences.

B. Final Thoughts on the Significance of the Cedar City Earthquake

As we reflect on the Cedar City earthquake, it becomes evident that its significance extends far beyond its immediate impact. While the absence of significant damage or injuries may lead some to dismiss it as inconsequential, the event serves as a sobering reminder of the ever-present threat posed by seismic activity. Moreover, it underscores the interconnectedness of natural processes and human vulnerability, highlighting the need for proactive measures to mitigate risks and enhance resilience.

The Cedar City earthquake also serves as a catalyst for reflection on the fragility of our built environment and the imperative of sustainable development practices. As communities continue to grapple with the dual challenges of urbanization and environmental change, the importance of integrating seismic risk considerations into planning and construction processes cannot be overstated. By prioritizing resilience and adaptability, we can build more robust communities capable of withstanding the test of time.

C. Consideration of the Broader Implications for Seismic Risk Management and Disaster Resilience

Looking beyond Cedar City, the lessons learned from this seismic event have broader implications for seismic risk management and disaster resilience worldwide. As populations continue to concentrate in urban centers situated in seismically active regions, the need for comprehensive risk assessment and mitigation strategies becomes increasingly urgent. By leveraging advances in technology and scientific understanding, we can enhance early warning systems, improve building codes, and strengthen infrastructure to minimize the impact of future earthquakes.

Moreover, the Cedar City earthquake underscores the importance of community engagement and empowerment in disaster preparedness and response efforts. By fostering collaboration between government agencies, non-profit organizations, and local residents, we can create more resilient communities capable of weathering the storm of natural disasters. From education initiatives to community drills, there are myriad opportunities to build a culture of preparedness and resilience from the ground up.

www.ingramcontent.com/pod-product-compliance
Lightning Source LLC
Chambersburg PA
CBHW070416230526
45471CB00006B/2840